Contents

Preface..3

Chapter 1: Origin of Life………………………………………….……..4-14

1.1 Millèr-Urey Experiment

1.2 Who was Stanley Miller?

1.3 Harold Clayton Urey

1.4 Oparin and Haldane Theory

1.5 Coacervate

1.6 Primordial Soup

1.7 Reddi Experiment

1.8 LazzaroSpallanzani

1.9 Discovery of Microorganisms

1.10 Fossil Evidence of Life

1.11 RNA World Hypothesis

1.12 What is Life?

Chapter 2: Biological Macromolecules……………………….…….15-35

2.1 Protein

2.2 Primary Structure of Proteins

2.3 Secondary Structure of Proteins

2.4 Tertiary Structure of Proteins

2.5 Carbohydrates

2.6 Fatty Acids

2.7 Nucleic Acids

2.8 Water

Chapter 3: Planets…………………………………………………………….36-43

6.1 Planets

Chapter 4: Life on Mars……………………………………………………..44-60

8.2 Meteorite

8.3 Various Missions on Mars

Chapter 5: Life on Europa…………………………………………..61-63

References

Preface

The first chapter describes Miller-Urey experiment, Oparin and Haldane Theory, Reddi experiment, fossils evidence of life, and RNA World hypothesis. The second chapter describes various biological macromolecules like carbohydrates, proteins, lipids, and nucleic acids. The third chapter planets and their composition. The fourth chapter describes various missions conducting on Mars. The fifth chapter describes life on Europa.

Chapter 1

Origin Of Life

Introduction:

Scientists are searching the questions "How life originate on the Earth?". Earth arose more than 3.5 billion years ago. Life arise from non-living matter. Whe life evolved on our Earth, very little amount of oxygen are found in the atmosphere.

1.1 Miller-Urey Experiment:

Miller-Urey experiment was supported by Alexander Oparin and J.B.S Haldane's (1952).

Requirment:

i. Water (H_2O).

ii. Methane (CH_4).

iii. Ammonia (NH_3).

iv. Hydrogen (H_2).

v. Two flasks.

Miller-Urey experiment was conducted by Stanley L. Miller and Harold C. Urey. The experiment was conducted in 1953 at the University of Chicago.

All the chemicals (CH_4, NH_3, H_2) were sealed inside glass tubes and flasks. Glass tubes and flasks are connected together in a loop. One flasks contains half filled water and another contains a pair of electrods. Then, the water was heated. Vapour released from water. The water vapour moves freely into the second flasks. Miller and Urey mixed the gasses and electrically sparked the mixture of gasses to produce lighting. Lighting was stimulated. The gasses were condensed into liquid.

Miller identified five types of amino acids like Glycine, α – Alanine, β – Alanine, Aspartic acid, and α – aminobutyric acid.

Result:

This theory proved that first life arise spontaneously from inorganic materials. All the inorganic molecules are combined to produce organic compounds.

1.2 Who was Stanley Miller?

Stanley Miller was born on March 7, 1930. He was an American chemist. He was born in California, Oakland, United States. Miller was considered the "Father of Prebiotic Chemistry". Miller was died on 20 May, 2007.

1.3 Harold Clayton Urey:

Harold Clayton Urey was born on April 29, 1893 in Walkerton, Indiana, United States. He was Nobel Prize Winner in Chemistry (1934). He discovered the deuterium. He contributing the Miller-Urey experiment.

1.4 Oparin and Haldane Theory:

Life arise from inorganic molecules. Oparin and Haldane thought that, inorganic molecules are reacted to produce amino acids and nucleotides. The first life arose from Ocean's Water.

1.5 Coacervate:

Coacervates mean aggregates of proteins, amino acids, and other hydrocarbons.

1.6 Primordial Soup:

The Primordial Soup Theory suggests that chemicals are combined to from amino acids. Amino acids are the building blocks of life. This Theory also suggests that life began on Earth Ocean Water at 3.8 billion years ago.

1.7 Reddi experiment:

We do not know when life originate on the Earth. Most scientists believed that life originate from non-living things such as maggots arose from wheat, lice from sweat, frog from dump mud, fish arose from mud. Several experiment have been conducted to disprove spontaneous generation.

Francesco Reddi is an Italian scientist. Francesco Reddi used to two different jars to this experiment. Both jars were exposed to the atmosphere. One jar was left open while another jar was covered with a cloth. Few days later, he observed the open jar contains maggots whereas the covered jar contains no maggots. Reddi concluded that maggots arose from eggs of the flies. He disproved the spontaneous generation.

1.8 LazzaroSpallanzani:

LazzaroSpallanzani was an Italian scientists. Lazzarospallanzani used four flasks for this experiment. Flask 1 was open and flask 2 was boiled and left open while flask 3 was sealed and flask 4 was boiled and sealed. After this experiment, he observed that falsk 1 and flask 2 was turn cloudy and microbes arises whereas flask 3 turn cloudy and microbes were found and flask 4 was did not turn cloudy and microbes not found.

1.9 Discovery of Microorganisms:

Earth is 4.6 billion years old. Scientist have evident that cells first appeared on Earth between 3.8 billion years ago; these organism were exclusively microbial. In fact microorganisms were the only life on Earth for most of its history. The discovery of the microbial world immediately raised question regarding the origin of microorganisms. Living organisms such as plants and animals do not originate spontaneously. However some belive that these microorganisms arose spontaneously and this theory came to be known as the theory of spontaneous generation or abiogenesis. The basic idea of spontaneous generation can easily be understood. For example, if food is allowed to stand for some time, it putrefies. When the putrefied material examined microscopically, it is found to be teeming with bacteria. Some people said they arose spontaneously from nonliving materials that is spontaneous generation. So, people had belived in generation- that living organism could develop from nonliving matter. But in 1748, the English priest John Needham (1713-1781) reported the results of his experiment on spontaneous generation. needham boiled mutton broth infused with plant or animal matter, hoping to kill all pre-existing microbes. He then sealed the flask. After a few days, Needham observed that the broth had become cloudy and contained microorganisms. He argued that the new microbes must have arisen spontaneously. Needham (1748) put the question to an experimental test. He wrote: "For my purpose therefore I took a quantity of mutton-gravy hot from the fire, and shut it up in a phial, closed with a cork so well masticated, that my precautions amounted to as much as if I had sealed my phial hermetically. I thus effectually excluded the exterior air, that it may not be said my moving bodies drew their origin from insects, or eggs floating in the atmosphere. I would not instill any water, lest, without giving it as intense a degree of heat, it might be thought these productions were conveyed through that element. My phials swarmed with life." Lazzarospallanzani (1729-1799) did not agree with Needham conclusions.

Today spontaneous generation is generally accepted to have been decisively dispelled during the 19th century by the experiment of Louis Pasteur. Pasteur prepared a nutrient broth similar to the broth one would use in soup. Then, he placed equal amount of the broth into two long nacked flasks. He left one flask with a straight neck. The other he bent to form an S shape. Then he boiled the broth in each flask to kill any living matter in the liquid. The sterile broths were then left to sit, at room temperature and exposed to the air, in their open mouthed flasks. After several weeks, Pasteur observed that the broth in the straight-neck flasks was discoloured and cloudy. Other broth in the curved neck flask had no changed. He concluded that germ in the air were able to fall unobstructed down the straight neck flask and contaminate the broth. Pasteur experiment showed that microbes cannot arise from nonliving materials under the conditions that existed on Earth during his lifetime. But his experiment did not proved that spontaneous generation never occurred. Another contribution of Louis pasteur to germ theory of disease. Germ theory of disease transmission was established by Pasteur. It states tha microorganisms known as pathogens or germs can lead to disease. Germs may refer to any type of microorganism that can cause disease such as protists, fungi, viruses, prions or viriods.

Pasteur is regarded the Father of bacteriology and pasteurization.

1.10 Fossil Evidence of Life:

Stromatolites are fossils. Stromatolites are formed by the growth of cyanobacteria.

1.11 RNA World hypothesis:

The RNA World Hypothesis was proposed by Carl Woese, Francis Crick, and Leslie Orgel. RNA can also transcribed by reverse transcription.

Alexander Rich was born on November 15, 1924. He was an American Biologist. He first gave the concept RNA World (1962).

Walter Gilbert was born on March 21, 1932 in Boston, Massachusetts, United States. He was an American molecular biologist. He first coined the term RNA World in 1986.

Many scientists believe that RNA first come on the Earth because RNA to store genetic information and catalyse chemical reactions.

1.12 What is Life?

There is no definition of life. Millions of years ago, life appears on the Earth. Life requires energy. Life may be reproduce, growth, and metabolise. Life consist of carbon, hydrogen, oxygen, sulphur, phosphorus and hydrogen. We belief that God are creates life to all living organisms. All living organisms are made up of cells. Cells are the fundamentals unit of life. Living thing consume nutrients. Living organisms may be reproduce. Reproduction may be asexual or sexual. Reproduction can create to produce new organism. So, life is capable of reproduction, metabolism, replication, and evolution.

Chapter 2

Biological Macromolecules

Biological macromolecules are carbohydrates, proteins, lipids and nucleic acids.

2.1 Protein:

Proteins are present in all living organisms. Protein consists of amino acid residues joined by peptide bonds. They are composed of one or more chains of amino acids. Amino acids are made up of carbon, hydrogen, nitrogen, oxygen, or sulphur. Amino acids are the building blocks of proteins.

Function of Proteins:

i. They catalyse metabolic reactions.

ii. They perform essentials for replicating DNA.

Types of Proteins:

i. Fibrous.

ii. Globular.

Iii. Membrane.

Fibrous:

Fibrous proteins are do not denature and generally insoluble in water. They also help in protection and support. Examples are Fibrous proteins are listed below-

- Fibronectin.
- Titin.
- Myosin.
- Spectrin.
- Collagen.
- Tropomysin.
- Elastin.
- Keratin.
- Tau
- Tubulin.

Globular Proteins:

Globular proteins are generally water soluble. Globular protein may be round or spherical. This proteins are help in transporting, regulating and catalysing. Examples of Globular proteins are given below-

- Albumins.
- Myoglobin.
- Ependymin.
- Selectin.
- Ig A
- Ig D
- Ig E
- Ig G
- Ig M
- Thrombin.
- Serum Albumin.

Membrane Protein:

Membrane Proteins are three types including integral membrane protein, peripheral membrane protein, anfdlipd anchored proteins. Examples of membrane proteins are given below-

- CFTR
- Glucose Transported.
- p^{53}
- Rhodopsin.
- Potassium channel.
- Myo D
- Hydrolases.
- Transferases.
- Scramblase.
- C-myc.

2.2 Primary Structure of Proteins:

Amino acids are link each other to forming proteins. Proteins are the backbone of amino acids. Primary structure of proteins are starting from the (N) terminal end and end to the (C) terminal end. Amino acid contains hydrogen atom, carboxyl group, amino group, and "R" group.

2.3 Secondary Structure of Proteins:

Secondary Structure of proteins are the α – helix and β- pleated sheet. The α-helix is right hand helix. α- helix are stabilized by hydrogen bonds between the NH and CO groups.

2.4 Tertiary Structure of Proteins:

Secondary structures are converted to form tertiary structure. Tertiary structures are three dimensional shape.

2.5 Carbohydrate:

Carbohydrates are consists of carbon, hydrogen, and oxygen. Carbohydrates are macronutrients. Carbohydrates are source of energy for the body. Carbohydrates are the building blocks of polysaccharides.

Types:

Carbohydrates are various types include- monosaccharide, disaccharide, and polysaccharide.

Monosaccharide:

Monosaccharides are the smallest sugar unit. Monosaccharides are colourless and dissolved in water. Glucose, Fructose and Galactose are monosaccharide.

Disaccharide:

When one monosaccharide are joined to another monosaccharide by glycosidic linkage to form disaccharides. Sucrose, maltose, and lactose are disaccharide.

Polysaccharide:
When many monosaccharides are joined to each other forming polysaccharide. Cellulose, Chitin, glycogen, and starch are polysaccharide.

Function of Carbohydrate:

i. Carbohydrate provide energy.

ii. Carbohydrate regulate the blood glucose.

iii. Carbohydrates are essentials for brain function.

2.6 Fattty Acids:

Fats are essentials part for health. We cannot live without fat. Fats are non-soluble in water. Fatty acids are the building blocks of fats. Fatty acids are soluble in water.

Types of Fatty Acids:

Fatty Acids are of two types-

i. Saturated Fatty Acids.

ii. Unsaturated Fatty Acids.

Saturated Fatty Acid:

Saturated Fatty acids do not contain double bonds.

Unsaturated Fatty Acids:

Unsaturated Fatty acids contain one or more double bonds.

2.6 Nucleic Acid:

Nucleic acids are building blocks of living organisms. Nucleic acids are macromolecules that transfers genetic information from one generation to another generation. Nucleic acids include deoxyribonucleic acid (DNA) and ribonucleic acid (RNA).

Nucleotides are consists are of three parts-

i. Nitrogenous base.

ii. A five carbon sugar.

iii. A phosphate group.

DNA:

DNA is double-stranded and composed of Adenine, Guanine, Cytosine, and Thymine.

RNA:
RNA is single stranded and composed of Adenine, Guanine, Cytosine, and Uracil.

2.7 Water:

Hydrogen and oxygen mixed with each other to from water. Wter as a solvent. Water is the essential part of our life. Wter are found all over the world. Water molecules are composed of two hydrogen atom linked on oxygen atom. Its chemical formula is H_2O. Water covers 71% of the Earth. Earth surface contains more amounts of water.

Traces of water are found on Mercury, Venus, Mars, and The Moon

Chapter 3

Planets

6.1 Planets:

The inner planets are Mercury, Venus, Earth, and Mars (terrestrial planets) while the outer planets are Jupiter, Saturn, Uranus, and Neptune (giant planets).

Mercury:

Mercury is the smallest planet and it has no satellite. Mercury is composed of atomic oxygen, sodium, magnesium, atomic hydrogen, potassium, calcium, helium, and trace amounts of iron, aluminium, argon, dinitrogen, dioxygen, carbon dioxide, water, vapour, xenon, krypton, neon. Mercury is the closet planet to the sun.

Venus:

Venus is the second planet from the sun but it has no satellite. Venus is composed of 96.5% Carbon dioxide, 3.5% Nitrogen, 0.015% sulphur dioxide, 0.0070% argon, 0.0020% water vapour, 0.0017% Carbon monoxide, 0.0012% Helium, 0.0007% Neon, and trace amounts of carbonyl sulphide, hydrogen chloride, hydrogen fluoride.

Earth:

We are all lives in the Earth. Earth is the third planet from the sun. Moon is the satellite of Earth. Earth is composed of 78.08% Nitrogen, 20.95% Oxygen, ~1% water vapour, 0.9340% argon, 0.0408% Carbon dioxide, neon, helium, methane, krypton, and hydrogen. Earth surface covered with 71% water.

Mars:

Mars is the fourth planet from the sun. Mars is called as the "Red Planet". Mars is composed of 95.97% Carbon dioxide, 1.93% Argon, 1.89% Nitrogen, 0.146% Oxygen, carbon monoxide, water vapour, nitrogen dioxide, neon, hydrogen deuterium oxide, krypton, and xenon.

Jupiter:

Jupiter is the largest planet in the solar system. Jupiter is composed of hydrogen, helium, methane, ammonia, hydrogen deuterium, ethane, water, and ammonium hydrosulfide.

Saturn:

Saturn is the second largest planet in the solar system. Saturn is composed of hydrogen, helium, methane, ammonia, hydrogen deuterium, ethane, water and ammonium hydrosulfide.

Uranus:

Uranus is composed of hydrogen, helium, methane, hydrogen sulfide, ammonia, water, ammonium hydrosulfide, and methane hydrate.

Chapter 4

Prospects For Life On Mars

8.1 Life on Mars:

Mars is the fourth planet from the sun. Average distance from sun to mars is 228 million km. The diameter of the Mars is 4,220 miles. Mars round the sun in 687 Earth day. This planet happens in various phenomenon like volcanoes, impacts, wind, crystal movements, and chemical reactions. Mars is made up of carbon dioxide (CO_2), argon (Ar), Nitrogen (N_2), trace amount of oxygen, and water vapour. Many spacecraft are sending to mars for searching life on Mars. The first successful mission was conducted in 1965 by Mariner 4 flyby. Mars is also called as "Red Planet." Mars soil and atmosphere look like red colour. Mars is rocky planet. We all know that, Mars cannot support life. But scientists are try to develop life on Mars. It is very challenging to develop life on Mars.

Mars was observed to telescope by Galileo Galilei in 1610. Due to its red colour appearance on the surface its visible to the naked eye. In today, Mars is very lonely planet. In today, there is no life on Mars. Scientists can try to develop life on Mars.

Past:

In Early Mars was very cold. In early time, Mars was colder than the Earth. In early time, volcanism have been happen. Water flows surface of Mars.

Curiosity was car sized rover operated by NASA. Curiosity was launched on November 26, 2011. Curiosity was landing on August 6, 2012. Objective of Curiosity was inventions of organic carbon compounds, investigate the building blocks of life.

According to NASA Rover team identified sulphur, nitrogen, hydrogen, oxygen, phosphorus, and possibly carbon. According to NASA reported from curiosity studying, Mars contain freshwater lake.

When we look at the Mars today for living. Scientists are found more evidences for life possible on Mars. Some evidences are given below:

Liquid water evidence:

Mars Reconnaissance Orbiter (MRO) is operated by NASA. Mars Reconnaissance Orbiter is launch on August 12, 2005.

Objective:

i. Study the Martian climate, geology, and weather.

ii. Search for sign of liquid water.

NASA Mars Reconnaissance Orbiter (MRO) is provide the evidence that liquid water flows on Mars.

Methane On Mars:

Methane present on Mars have been reported in various mission.

Mars express was operated by European Space Agency (ESA). Mars express was launch on 2 June, 2003. The Planetary Fourier Spectrometer instrument is currently used by European Space Agency. Professor Vittorio Formisano on March 2004 suggests that presence of Methane on Mars.

Geysers:

A Geyser is rare phenomenon occurs in few places. Geyser is a discharge of water by convection. Hot water rises the top by pressure. Geysers are occur in few places in Earth e.g. Yellowstone National Park, United States, Valley of Geysers, Russia, El Tatio, Chile, Taupo Volcanic Zone, New Zealand, Iceland. Geysers can be happen in atmospheric pressure, dry venting, water driven erosion, and geothermal activity.

Scientists proposed that Geysers are happen in Mars.

8.2 Meteorite:

ALH84001:

ALH84001 is a meteorite. ALH84001 was found on December 27, 1981 in Antartica. Scientists are thought to ALH84001 have originated on Mars. ALH84001 are 20-100 nanometers.

ALH84001 have come from Mars.

Nakhla Meteorite:

Nakhla is a meteorite . The Nakhla is found in the Abu Hommos district, Alexandria Governorate, Khedivate of Egypt. Scientists suggests that Nakhla contains carbonates and hydrous minerals and various amino acids like aspartic acid, glutamic acid, glycine, alanine, and γ- aminobutyric acid.

Shergotty:

Shergotty meteorite was found in the Gaya district, Bihar on 25 August, 1865. It consists of pyroxene.

Yamato 000593:

Yamato 000593 is a meteorite. This meteorite was found in Antartica. Yamato 000593 was composed on pyroxene (85%), and olivine (10%).

8.3 Various Mission on Mars:

Scientista are try to finding life on Mars. Scientists are made many spacecraft. Spacecraft are sends to Mars to search for life, obtain high resolution images, and discover new things. The average distance from Earth to Mars is 36 million miles.

Viking 1:

Viking is a spacecraft. Viking is consists of two main parts- orbiter and lander. Viking 1 orbiter consists of two vidicon cameras for capturing photograph.

Orbiter capture the photograph on Mars and lander designed to study the Mars.

Viking was manufactured by Jet Propulsinon Laboratory /Martin Marietta.

Viking 1 was launched on August 20, 1975. Viking was operated by NASA and send to Mars. Viking 1 was arrived into Mars on June 19, 1976.

Viking 1 carried biological experiments for evidences of life like pyrolytic release experiment (PR), the labelled release experiment, and gas exchange experiment and organic experiment like gas chromatograph-mass spectrometer (GC/MS).

Viking 2:

Viking 2 was manufacture by Jet Propulsion Laboratory/Martin Marietta. Viking 2 was launch on September 9, 1975. Viking 2 arrived into Mars on August 7, 1976.

Viking 2 carried soil experiment on Mars. Viking 2 experiment on tested soil contains silicon, iron, magnesium, aluminium, sulphur, calcium, titanium.

Viking 2 carried biological experiments for evidences of life like pyrolytic release experiment (PR), the labelled release experiment, and gas exchange experiment and organic experiment like gas chromatograph-mass spectrometer (GC/MS).

Viking Experiments:

Pyrolytic Release Experiment:

The Pyrolytic release experiment (PR) was carried by Viking mission. Soil samples are added to radioactively labelled CO_2, CO, and Krypton. Samples are incubated.

Result:

Life synthesizes organic materials from Carbon dioxide (CO_2) and carbon monoxide (CO).

Labeled Release Experiment:

Soil samples are added to the nutrient, He gas and incubate the sample.

Result:

i. The presence of microbial activity in the soil.

ii. Metabolize simple organic comnpounds.

iii. Produce gaseous wastes.

Gas Exchange Experiment:

Soil are incubated with nutrient medium. First, Chicken soup added to the soil. Then incunbate in the Mars atmosphere.

Result:

Gas emitted from this experiment. It indicates that organism coming to life.

Mars 1:

Mars 1 was launch in 1962.

Mars 2 and Mars 3:

Mars 2 was spacecraft. Mars 2 and Mars 3 was operated by Soviet Union. The Mars 2 and Mars 3 was launch on May 28, 1971. The Mars 2 and Mars 3 consists of orbiter and lander.

Objective:

i. Study the topography of the martian surface.

ii. Properties of the atmosphere.

Mariner 4:

Mariner 4 was manufactured in Jet Propulsion Laboratory. Mariner 4 was launched on November 28, 1964. Mariner 4 was operated by NASA.

Mariner 4 is as a spacecraft. This mission was in 3 years, 23 days in duration. Objective of this mission was observation of Mars and transmits these observation to Earth.

Mariner 6 and Mariner 7:
The Mariner 6 and Mariner 7 was operated by NASA/JPL. The Mariner 6 and Mariner 7 was launch on February 25, 1969. The Mariner 6 and Mariner 7 was manufactured by Jet Propulsion Laboratory.

Objective:

i. Search for extraterrestrial life.

Ii. Study the surface and atmosphere of Mars.

Mariner 9:

Mariner 9 was launch on May 30, 1971. Mariner 9 was operated and manufactured by NASA/JPL.

Mars Global Surveyor:

Mars Global Surveyor was launch on 7 November, 1996.

Objective:

i. Characterize the surface features on Mars.

Pheonix:

Pheonix was a spacecraft. It was manufactured by Lockheed Martin Space systems. It was launched on August 4, 2007.

Objective:

i. Study the geological history of water.

Dawn:

Dawn was launched by NASA ON September 27, 2007. Dawn was manufactured by orbital sciences/ JPL/NASA.

Spirit:

Spirit was operated by NASA. Spirit was launched on June 10, 2003.

Objective:

i. Search for different types of rock and soil.

Rosetta:

Rosetta was a spacecraft. Rosetta was launch on 2 March, 2004. Rosetta was manufactured by Astrium.

Objective:

i. Search for organic compounds.

Mars Pathfinder:

Mars Pathfinder was operated by NASA and Jet Propulsion Laboratory. Mars pathfinder was an robotic spacecraft. Mars Pathfinder was launch on December 4, 1996. Mars pathfinder was landing on July 4, 1997.

Objective:

The objective of this mission was to send scientific instrument to Mars. This spacrcraft send 16500 pictures.

Sojourner:

Sojourner was operated by NASA. Sojourner was launch on December 6, 1996. It was a successful mission.

Mars Sample-Return Mission:

A Mars sample return mission is a very important mission to spacecraft enters into Mars. After reaching on Mars, collect rock and dust sample on Mars and return into Earth.

ExoMars:

Exo Mars is operated by European Space Agency/Russian Space Research Institute.

Objective:

i. To search past martian life.

Mars Orbiter Mission:

The Mars Orbiter Mission was operated by Indian Space Research Organozation (ISRO). Mars Orbiter Mission was launched on 5 November, 2013. Mission Orbiter Mission was manufactured by ISAC. Mars Orbiter Mission was also called as Mangalyaan.

Objective:

i. Study the morphology, topography, and martian atmosphere of Mars.

Oppurtunity:

Opportunity was a robotic rover. Oppurtunity was launch on July 7, 2003. Oppurtunity was landing on January 25, 2004.

Objective:

The objective of this mission was search for varieties of rock on the Mars surface.

ii. Search for iron containing minerals.

Mars One Mission:

The Mars One Mission is a planning to going on Mars in 2032. So, One day we go to Mars. This mission is one way mission. No return from Mars to Earth and permanent settlement on Mars. According to this Mission, Human going to Mars in 2031 and landing on Mars in 2032. Every two year additional crews were send to Mars. So, Mars One Mission is the one way mission to the red planet.

Bas Landsdorp born on 5 March, 1977. He received his M.Sc in Mechanical engineering from the University of Twente in 2003. Bas Landsdorp is the Co-founder and CEO of Mars One.

Spacecraft	Launch Date
Mariner 4	28 November, 1964
Mariner 6	25 Febuary, 1969
Mariner 7	27 March, 1969
Mars 2	19 May, 1971
Mars 3	28 May ,1971
Mariner 9	30 May, 1971
Viking 1	20 August, 1975
Viking 2	9 September, 1975
Mars Global Surveyor	7 November, 1996
Mars Pathfinder	4 December, 1996
Sojourner	6 December, 1996
Spirit	10 June, 2003
Oppurtunity	8 July, 2003
Rosetta	2 March, 2004
Pheonix	4 August, 2007
Dawn	27 September, 2007
Marco	5 May, 2018

Conclusion:

Did life ever evolved on Mars? NASA (National Aeronautics and Space Administration) search for life on Mars. NASA conducts various mission for knowing past or present life on Mars.

Chapter 5

Life on Europa

Europa was discovered by Galileo Galilei. The volume of Europa is 1.593×10^{10} km^3. Europa is made up of iron-nickel core, silicate rock, and water-ice crust. Researchers search in life on Europa. Europa is orbits of Jupiter. According to Voyager spacecraft in 1979, Europa must contains liquid water. Astrobiologists believed that, Europa contains abundant water. Equtorial diameter of Europa is 1940 milles. Europa is the good place in our solar system. Europa orbits Jupiter every 3.5 Earth days. Europa has no moons and no rings. The Europa surface found is mostly solid water-ice. The Europa surface is also consists of extremely thin oxygen.

Europa was discovered by Galileo Galilei in 1610. Europa was discovered in 8 January 1610. Alternatives name of Europa is Jupiter II. Europa is the satellite of Jupiter. Europa is the sixth largest moon in the solar system. Europa is smaller than moon.

Evidences for Life:

Water:

Water is the basic needed for life. Without water life cannot possible. According to dada, Europa may have plenty of water.

Chemistry:

Life requires various types of essentials chemical ingredients like carbon, hydrogen, nitrogen, oxygen, phosphorus, and sulphur.

Energy:

Energy is the capacity to do work. Life requires energy for survival.

Exploration of Europa:

Europa clipper is the mission of NASA. The main aim to this mission is search for signs of potential habitability of Europa. This mission is launch on probably in 2020.

JUICE is the mission of European Space Agency. This mission was launch on Probably 2022.

References:

1. Miller-Urey Experiment - Amino Acids & Origins of Life on Earth, www.juliantrubin.com/bigten/miller_urey_experiment.html.

2. "A Brief Explanation Of Miller Urey Experiment". BYJUS, 2020, https://byjus.com/biology/miller-urey-experiment/.

3. "Miller–Urey Experiment". En.Wikipedia.Org, 2020, https://en.wikipedia.org/wiki/Miller–Urey_experiment.

4. http://phoenix.lpl.arizona.edu/mars145.php

5. https://study.com/academy/lesson/stanley-miller-theory-experiment-apparatus.html

6. https://en.wikipedia.org/wiki/Stanley_Miller#cite_note-23

7. Reece, Jane B., and Neil A. Campbell. Campbell Biology. Harlow: Pearson Education, 2011. Print.

8. IBM TJ Watson Researcher CenterIsidoreRigoutsos Manager Bioinformatics Group, Gregory Stephanopoulos Professor of Chemical Engineering and Biotechnology MIT (2006), Systems.

9. Biology: Volume I: Genomics, Oxford University Press, p. 6.

10. "Origin Of Life: Twentieth Century Landmarks." Origin Of Life: Oparin-Haldane Hypothesis. N.p., n.d. Web. 28 Oct. 2012. <http://www.simsoup.info/Origin_Landmarks_Oparin_Haldane.html>.

11. "The Miller/Urey Experiment." The Miller/Urey Experiment.N.p., n.d. Web. 28 Oct. 2012. <http://www.chem.duke.edu/~jds/cruise_chem/Exobiology/miller.html>.

12. Reece, Jane B., and Neil A. Campbell. Campbell Biology. Harlow: Pearson Education, 2011. Print.

"Geysers and Volcanic Activity."*Geyser | Geysers and Volcanic Activity*, universe.dk/en/bliv-klog/geyser.

13. "Experimental Design." *Mit*, web.mit.edu/12.000/www/finalpresentation/experiments/pr.html.

14. Carr, Dr. Michael H. and Benton Clark et al. "An Exobiology Strategy for Mars Exploration."
Online: http://cmex.arc.nasa.gov/MarsNews/mars_papers/Docs/mars_strat.html

15. NASA, "Science on Mars."
Online: http://history.nasa.gov/SP-4212/ch11-5.html

16. Caplinger, Michael. "Life on Mars."
Online: http://barsoom.msss.com/http/ps/life/life.html

17. For a handy semi-animation of this experiment, go to http://cybooks.com/ahill3/hill03.html

18. "Overview." *NASA*, NASA, solarsystem.nasa.gov/planets/mars/overview/.

19. "Mars: Facts, Information, History& Definition." *The Nine Planets*, 22 Nov. 2019, nineplanets.org/mars/.

20. "Mars Facts." *NASA*, NASA, 27 July 2019, mars.nasa.gov/all-about-mars/facts/.

21. "Experimental Design." *Mit*, web.mit.edu/12.000/www/finalpresentation/experiments/gx.html.

22. Mars from orbit - 2, Spaceflight, Unnumbered, 118-120, Mar. 1972.

23. Levin, Gilbert V. and Patricia Ann Straat, "Recent Results From the Viking Labeled Release Experiment on Mars." Journal of Geophysical Research. Vol. 82, No. 28 September 30, 1977

24. Smolders, P., Results of Soviet research into Mars, Interavia, 27, 1139-1141, Oct. 1972.

25. Harvey, B., The new Russian space programme from competition to collaboration, John Wiley & Sons, Chichester, England, 1996.

26. Johnson, N. L., Handbook of soviet lunar and planetary exploration - volume 47 science and technology series, Amer. Astronau. Soc. Publ., 1979.

27. Oja, H., Soviet Mars landers, Spaceflight, 15, No. 7, 242-245, July 1973.

28. Perminov, V. G., The difficult road to Mars - A brief history of Mars exploration in the Soviet Union, NASA, No. 15, Wash, DC, July 1999.

29. "Biosignature." *Wikipedia*, Wikimedia Foundation, 30 Nov. 2019, en.wikipedia.org/wiki/Biosignature.

30. "Mars Orbiter Mission." *Wikipedia*, Wikimedia Foundation, 30 Jan. 2020, en.wikipedia.org/wiki/Mars_Orbiter_Mission.

31. "Biosignature." *Biosignature Definition*, web.archive.org/web/20100316062919/www.sciencedictionary.com/definition/biosignature.html.

32. "Mars Sample-Return Mission." *Wikipedia*, Wikimedia Foundation, 15 Dec. 2019, en.wikipedia.org/wiki/Mars_sample-return_mission.

33. https://spaceplace.nasa.gov/asteroid/en/

34. https://lambda.gsfc.nasa.gov/product/suborbit/POLAR/cmb.physics.wisc.edu/tutorial/olbers.html

35. Ap. J. 367, 399 (1991). The author, Paul Wesson, is said to be on a personal crusade to end the confusion surrounding Olbers' paradox.

36. Darkness at Night: A Riddle of the Universe, Edward Harrison, Harvard University Press, 1987http://hyperphysics.phy-astr.gsu.edu/hbase/Astro/olbers.html

37. https://www.google.com/amp/s/www.space.com/amp/51-asteroids-formation-discovery-and-exploration.html

38. https://www.infoplease.com/math-science/biology/genetics-evolution/origin-of-life-spontaneous-generation

39. https://www.nasa.gov/press-release/nasa-confirms-evidence-that-liquid-water-flows-on-today-s-mars

40. https://solarsystem.nasa.gov/moons/saturn-moons/titan/in-depth/

41. https://en.m.wikipedia.org/wiki/Life_on_Titan

42. https://www.khanacademy.org/science/high-school-biology/hs-biology-foundations/hs-biology-and-the-scientific-method/a/what-is-life

43. https://en.m.wikipedia.org/wiki/Mars_Pathfinder

44. https://en.m.wikipedia.org/wiki/Opportunity_(rover)

45. https://www.thoughtco.com/nucleic-acids-373552

46. https://www.infoplease.com/encyclopedia/science/biology/concepts/exobiology

47. Bernal, J.D. (1949) The physical basis of life. Proc. of the Physical Society.62, 10, 358B.

48. https://www.google.com/amp/s/www.space.com/amp/25219-drake-equation.html

49. https://en.m.wikipedia.org/wiki/Drake_equation

50. https://www.google.com/amp/s/www.space.com/amp/25325-fermi-paradox.html

51. https://www.google.com/amp/s/amp.interestingengineering.com/a-brief-history-of-the-telescope-from-1608-to-gamma-rays

52. https://en.m.wikipedia.org/wiki/Earth

53. https://europa.nasa.gov/europa/life-ingredients/

54. Gould, S.J. (1989) The iconography of an expectation. In Wonderful Life, W.W. Norton, New York,

pp. 23–52.

55. Lineweaver, C.H. (2005), Book Review of "Intelligent Life in the Universe: From Common

56. https://en.m.wikipedia.org/wiki/Bas_Lansdorp

57. https://en.m.wikipedia.org/wiki/Mars_One

58. Origins to the Future of Humanity" by Peter Ulmschneider. Astrobiology, Volume 5, Number 5

59. Lineweaver, C.H. (2006), We Have Not Detected Extraterrestrial Life, or Have We? in ``Life As We Know It'' edt. J. Seckbach, Vol. 10 of a series on Cellular Origin, Life in Extreme Habitats and

60. Astrobiology, Springer, Dordrecht pp 445-457, ISBN 1-4020-4394-5

61. Lineweaver, C.H., (2009), Paleontological Tests: Human-like Intelligence is not a Convergent Feature of Evolution in "From Fossils to Astrobiology", edt. J. Seckbach& M. Walsh Vol 13 of a series on

62. Cellular Origins and Life in Extreme Habitats and Astrobiology, Springer, pp 353-368

63. Mayr, E. (1994) Does it pay to acquire high intelligence? Perspect. Biol. Med. 37(3), 337–338.

64. Mayr, E. (1995) Can SETI succeed? Not likely. Bioastron.News 7(3). Available online at:

http://seti.planetary.org/Contact/debate/Mayr.htm.

65. Prigogine, I. (1980), From Being to Becoming Freeman, NY.

Sagan, C. (1995) The abundance of life-bearing planets. Bioastron.News 7(4). Available online at:

66. http://seti.planetary.org/Contact/debate/Sagan.htm.

67. Schrödinger, E. (1944), What is life? and other scientific essays Double Day Anchor, Garden City, NY.

68. Simpson, G.G. (1964) The non-prevalence of humanoids. Science 143, 769–775.

69. https://examples.yourdictionary.com/examples-of-protein.html

70. https://www-livescience-com.cdn.ampproject.org/v/s/www.livescience.com/amp/51976-carbohydrates.html?amp_js_v=a2&amp_gsa=1&usqp=mq331

AQCKAE%3D#aoh=15795924655305&amp_ct=1579592472398&referrer=https%3A%2F%2Fwww.google.com&amp_tf=From%20%251%24s&ampshare=https%3A%2F%2Fwww.livescience.com%2F51976-carbohydrates.html

71. http://www.bbc.com/earth/story/20170101-there-are-over-100-definitions-for-life-and-all-are-wrong

72. https://www.medicalnewstoday.com/articles/161547.php#what-are-carbohydrates

73. https://study.com/academy/lesson/monosaccharides-definition-structure-examples.html

74. https://biologydictionary.net/polysaccharide/

75. https://www.rsc.org/Education/Teachers/Resources/cfb/carbohydrates.htm

76. https://www.skillsyouneed.com/ps/fat.html

77. https://en.m.wikipedia.org/wiki/Fatty_acid

78. https://kidshealth.org/en/parents/fatty-acids.html

79. https://www.britannica.com/science/water

80. Carr, M. H. Water on Mars. London, U.K.: Oxford University Press, 1996.

81. Chang, S. "The Planetary Setting of Prebiotic Evolution." In Early Life on Earth, ed. S. Bengston. New York: Columbia University Press, 1994.

82. Klein, H. P. "The Search for Life on Mars: What We Learned from Viking." Journal of Geophysical Research. 103 (1998):28463–28466.

83. Lemonick, M. D. Other Worlds: the Search for Life in the Universe. New York: Simon and Schuster, 1998.

84. Malin, M. C., and K. S. Edgett. "Evidence for Recent Groundwater Seepage and Surface Runoff on Mars." Science 288 (2000):2330–2335.

85. Pace, N. R. "A Molecular View of Microbial Diversity and the Biosphere." Science 276 (1997):734–740.

86. http://www.waterencyclopedia.com/A-Bi/Astrobiology-Water-and-the-Potential-for-Extraterrestrial-Life.html#ixzz6BegU5Pu4

87. http://www1.lsbu.ac.uk/water/astrobiology.html

88. https://www.esa.int/Science_Exploration/Human_and_Robotic_Exploration/Exploration/Extraterrestrial_life

89. https://www.ck12.org/book/cbse_biology_book_class_xii/section/9.3/

90. Harwood, R. (2012). Patterns in palaeontology: The first 3 billion years of evolution. Palaeontology, 2(11), 1-22. Retrieved from http://www.palaeontologyonline.com/articles/2012/patterns-in-palaeontology-the-first-3-billion-years-of-evolution/.

91. Wacey, D., Kilburn, M. R., Saunders, M., Cliff, J., and Brasier, M. D. (2011). Microfossils of sulphur-metabolizing cells in 3.4-billion-year-old rocks of Western Australia. Nature Geoscience, 4, 698-702. http://dx.doi.org/10.1038/ngeo1238.

92. Primordial soup. (2016, January 20). Retrieved May 22, 2016 from Wikipedia: https://en.wikipedia.org/wiki/Primordial_soup.

93. Gordon-Smith, C. (2003). The Oparin-Haldane hypothesis. In Origin of life: Twentieth century landmarks. Retrieved from http://www.simsoup.info/Origin_Landmarks_Oparin_Haldane.html.

94. The Oparin-Haldane hypothesis. (2015, June 14). In Structural biochemistry. Retrieved May 22, 2016 from Wikibooks: https://en.wikibooks.org/wiki/Structural_Biochemistry/The_Oparin-Haldane_Hypothesis.

95. Kimball, J. W. (2015, May 17). Miller's experiment.In Kimball's biology pages.Retrieved from http://www.biology-pages.info/A/AbioticSynthesis.html#Miller's_Experiment.

96. Earth's early atmosphere. (Dec 2, 2011). In Astrobiology Magazine.Retrieved from http://www.astrobio.net/topic/solar-system/earth/geology/earths-early-atmosphere/.

97. McCollom, T. M. (2013). Miller-Urey and beyond: What have learned about prebiotic organic synthesis reactions in the past 60 years? Annual Review of Earth and Planetary Sciences, 41_, 207-229. http://dx.doi.org/10.1146/annurev-earth-040610-133457.

98. Powner, M. W., Gerland, B., and Sutherland, J. D. (2009). Synthesis of activated pyrimidine ribonucleotides in prebiotically plausible conditions. Nature, 459, 239-242. http://dx.doi.org/10.1038/nature08013.

99. Lurquin, P. F. (June 5, 2003). Proteins and metabolism first: The iron-sulfur world. In The origins of life and the universe (pp. 110-111). New York, NY: Columbia University Press.

100. Ferris, J. P. (2006). Montmorillonite-catalysed formation of RNA oligomers: The possible role of catalysis in the origins of life. Philos. Trans. R. Soc. Lond. B. Bio.l Sci., 361(1474), 1777–1786. http://dx.doi.org/10.1098/rstb.2006.1903.

101. Kimball, J. W. (2015, May 17). Assembling polymers.In Kimball's biology pages. Retrieved from

102. Montmorillonite. (2016, 28 March). Retrieved May 22, 2016 from Wikipedia: https://en.wikipedia.org/wiki/Montmorillonite.

103. Wilkin, D. and Akre, B. (2016, March 23). First organic molecules - Advanced. In CK-12 biology advanced concepts. Retrieved from http://www.ck12.org/book/CK-12-Biology-Advanced-Concepts/section/10.8/.

104. Hollenstein, M. (2015). DNA catalysis: The chemical repertoire of DNAzymes. Molecules, 20(11), 20777–20804. http://dx.doi.org/10.3390/molecules201119730.

105. Breaker, R. R. and Joyce, G. F. (2014). The expanding view of RNA and DNA function.Chemistry & biology, 21(9), 1059–1065. http://dx.doi.org/10.1016/j.chembiol.2014.07.008.

106. Alberts, B., Johnson, A., Lewis, J., Raff, M., Roberts, K., and Walter, P. (2002). A pre-RNA world probably predates the RNA world. In Molecular biology of the cell (4th ed.). New York, NY: Garland Science. Retrieved from http://www.ncbi.nlm.nih.gov/books/NBK26876/#_A1124_.

107. Moran, L. A. (2009, May 15). Metabolism first and the origin of life. In Sandwalk: Strolling with a skeptical biochemist. Retrieved from http://www.simsoup.info/Origin_Landmarks_Oparin_Haldane.html.

108. Kimball, J. W. (2015, May 17). The first cell?In Kimball's biology pages. Retrieved from http://www.biology-pages.info/A/AbioticSynthesis.html#TheFirstCell?

109. Kimball, J. W. (2015, May 17). Molecules from outer space?In Kimball's biology pages. Retrieved from http://www.biology-pages.info/A/AbioticSynthesis.html#Molecules_from_outer_space?.

110. Jeffs, W. (2006, November 30). NASA scientists find primordial organic matter in meteorite. In NASA news.Retrieved from http://www.nasa.gov/centers/johnson/news/releases/2006/J06-103.html.

111. Alberts, B., Johnson, A., Lewis, J., Raff, M., Roberts, K., and Walter, P. (2002). The RNA world and the origins of life. In Molecular biology of the cell (4th ed.). New York, NY: Garland Science. Retrieved from http://www.ncbi.nlm.nih.gov/books/NBK26876/.

112. Branch of Isotope Geology, United States Geological Survey. (2014, October 31). The age of the Earth. In USGS geology in the parks. Retrieved from http://geomaps.wr.usgs.gov/parks/gtime/ageofearth.html.

113. Ferris, J. P. (2005). Mineral catalysis and prebiotic synthesis: Montmorillonite-catalyzed formation of RNA. Elements, 1, 145-149. Retrieved http://dx.doi.org/10.2113/gselements.1.3.145.

114. Hashizume, H. (2012). Role of clay minerals in chemical evolution and the origins of life. In Valaškova, M. and Martynkova, G. (Eds.), Clay minerals in nature - Their characterization, modification, and application. InTech. http://dx.doi.org/10.5772/50172.

115. Lattimer, J. M. (2016). Chemical evolution theory of life's origins. In AST 248: The search for life in the universe. Retrieved from http://www.astro.sunysb.edu/lattimer/AST248/lecture_13.pdf.

116. Lovgren, S. (2003, October 2). Did comets make life on Earth possible? In National geographic news. Retrieved from http://news.nationalgeographic.com/news/2003/10/1002_031002_cometstudy.html.

117. Matson, J. (2010, February 15). Meteorite that fell in 1969 still revealing secrets of the early solar system. In Scientific American. Retrieved from http://www.scientificamerican.com/article/murchison-meteorite/.

118. Miller, S. L. (1953). A production of amino acids under possible primitive Earth conditions. Science, 117(3046), 528-529. http://dx.doi.org/ 10.1126/science.117.3046.528.

119. Moritz, A. (2010, July 1). The origin of life.In The TalkOrigins archive.Retrieved from http://www.talkorigins.org/faqs/abioprob/originoflife.html.

120. Murchison meteorite (April 10, 2016). Retrieved April 15, 2016 from Wikipedia: https://en.wikipedia.org/wiki/Murchison_meteorite.

121. Newman, W. (2007, July 9). Age of the earth.In Geologic time.Retrieved from http://pubs.usgs.gov/gip/geotime/age.html.

122. Pablo. (2009). The origin of life, part II: Polymerization. In Pablo's origins blog.Retrieved from http://pablosorigins.blogspot.com/2009/11/origin-of-life-part-ii.html.

123. Primordial soup. (2016, January 20). Retrieved May 22, 2016 from Wikipedia: https://en.wikipedia.org/wiki/Primordial_soup.

124. Purves, W. K., Sadava, D., Orians, G. H., and Heller, H. C. (2003). Theories of the origin of life. In Life: The science of biology (7th ed., pp. 35-37). Sunderland, MA: Sinauer Associates, Inc.

125. Raven, P. H. and Johnson, G. B. (2002). There are many ideas about the origin of life. In Biology (6th ed., pp. 62-65). Boston, MA: McGraw-Hill.

126. Reece, J. B., Urry, L. A., Cain, M. L., Wasserman, S. A., Minorsky, P. V., and Jackson, R. B. (2011). Conditions on early Earth made the origin of life possible. In Campbell biology (10th ed., pp. 520-522). San Francisco, CA: Pearson.

127. Schultz, C. (2014, May 16). How do we know the Earth is 4.6 billion years old? In Smithsonian.com.Retrieved from http://www.smithsonianmag.com/smart-news/how-do-we-know-earth-46-billion-years-old-180951483/?no-ist.

128. Shapiro, R. (2007, February 12). A simpler origin for life.In Scientific American.Retrieved from http://www.scientificamerican.com/article/a-simpler-origin-for-life/.

129. Sczepanski, J. T. and Joyce, J. F. (2014). A cross-chiral RNA polymerase ribozyme. Nature, 515(7527), 440-442. http://dx.doi.org/10.1038/nature13900. Retrieved from http://www.ncbi.nlm.nih.gov/pmc/articles/PMC4239201/.

130. Service, R. F. (March 16, 2015). Researchers may have solved origin-of-life conundrum. In Science Magazine.Retrieved from http://www.sciencemag.org/news/2015/03/researchers-may-have-solved-origin-life-conundrum.

131. Sidney W. Fox (2016, March 10). Retrieved April 10, 2016 from Wikipedia: https://en.wikipedia.org/wiki/Sidney_W._Fox.

132. Speer, B. R. (1997). Cyanobacteria: Fossil record. In University of California Museum of Paleontology.Retrieved from http://www.ucmp.berkeley.edu/bacteria/cyanofr.html.

133. Stromatolite. (2016, May 12). Retrieved April 15, 2016 from Wikipedia: https://en.wikipedia.org/wiki/Stromatolite.

134. Stromatolites of Shark Bay. (2016). In Welcome to Shark Bay. Retrieved from http://www.sharkbay.org.au/places-to-visit/hamelin_pool_stromatolites.aspx

135. The Oparin-Haldane theory of the origin of life. (n.d.).In The origin of life on Earth.Retrieved from http://www.chem.ox.ac.uk/vrchemistry/chapter26/page11.htm.

136. University of California Museum of Paleontology. (2016.). The RNA World. In Understanding evolution.Retrieved from http://evolution.berkeley.edu/evolibrary/article/ellington_03.

137. University of California Museum of Paleontology. (2016). When did life originate? In Understanding evolution.Retrieved from http://evolution.berkeley.edu/evolibrary/article/0_0_0/origsoflife_02.

138. Wilkin, D. and Akre, B. (2016, March 23). First organic molecules - Advanced. In CK-12 biology advanced concepts. Retrieved from http://www.ck12.org/book/CK-12-Biology-Advanced-Concepts/section/10.8/.

139. https://en.m.wikipedia.org/wiki/Harold_Urey

140. https://www.windows2universe.org/earth/Life/coacervates.html

141. http://leiwenwu.tripod.com/primordials.htm

142. http://exploringorigins.org/index.html

143. http://www.ncbi.nlm.nih.gov/books/NBK26876/

144. http://www.nature.com/nrg/journal/v16/n1/full/nrg3841.html

145. http://www.evolutionnews.org/2015/06/on_the_origin_o_7097191.html

146. http://www.panspermia.org/rnaworld.htm

147. http://exploringorigins.org/ribozymes.html

148. https://spaceplace.nasa.gov/all-about-exoplanets/en/

149. https://solarsystem.nasa.gov/moons/jupiter-moons/europa/in-depth/

150. https://en.wikipedia.org/wiki/Europa_(moon)

151. https://solarsystem.nasa.gov/moons/jupiter-moons/europa/overview/

152. https://solarsystem.nasa.gov/missions/europa-clipper/in-depth/

153. https://solarsystem.nasa.gov/missions/juice/in-depth/

www.ingramcontent.com/pod-product-compliance
Lightning Source LLC
Chambersburg PA
CBHW040329220526
45473CB00009B/2624